Toni Taylor (at Kansas Rural Center) Newton KS Nov. 2, 2013

(she's now in her 90's)

KANSAS BARNS II

Sketches by

Martha Knudsen

Printed by
Taylor Printing, Inc.
Pratt, Kansas
1999

book I (out of print)
also a book III

4028 NE 48th Street
Newton, Kansas 67114

ISBN 0-9638978-1-0

Printed in the United States of America

FOREWORD

I would like to thank all the folks whose barns I used in my drawings and all the people who told me how to find so many of them.

Many of the people were able to provide me with histories of their barns through personal letters, newspaper clippings, or over the phone. A thank you also to the Marion, Butler and Chase County Historical Societies for their help.

I would further like to thank my family members for their support, especially my daughter, Martie, who put all the histories on computer for me.

Between the two books, I now have at least one barn featured from every county in Kansas. I know I have missed some very interesting barns behind trees and hills, maybe some day I'll be able to find them with help.

If you have time, be sure and take the back roads, if not to see barns, to see the countryside and small towns. We truly have a beautiful state and many interesting sites to see.

TABLE OF CONTENTS

NORTHWEST KANSAS:

TABLE OF CONTENTS Continued

NORTHEAST KANSAS:

TABLE OF CONTENTS Continued

NORTHEAST KANSAS Continued:

SOUTHWEST KANSAS:

TABLE OF CONTENTS Continued

SOUTHWEST KANSAS Continued:

SOUTHEAST KANSAS:

TABLE OF CONTENTS Continued

SOUTHCENTRAL KANSAS:

TABLE OF CONTENTS Continued

SOUTHCENTRAL KANSAS Continued:

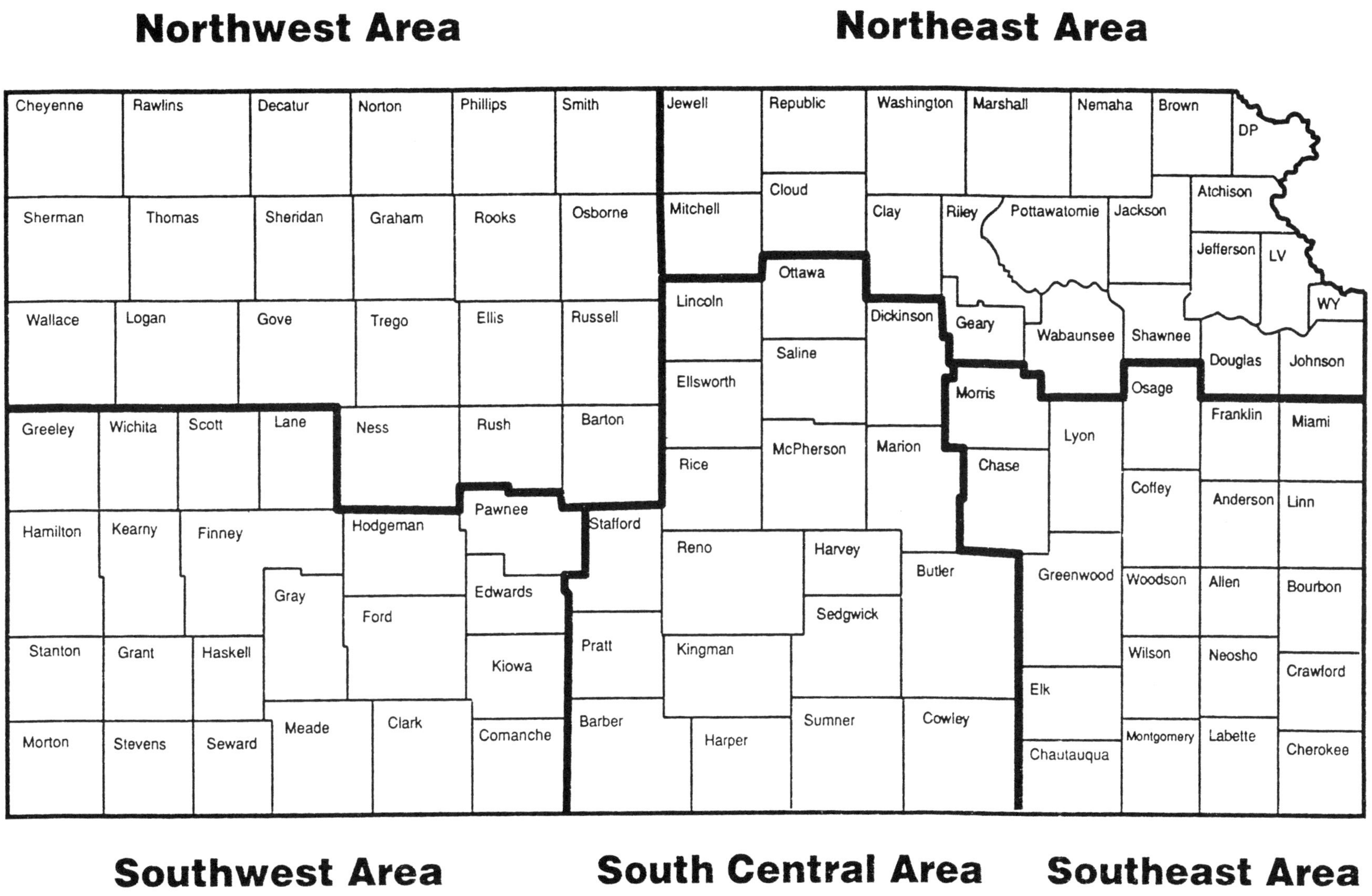
Northwest Area
Northeast Area
Cheyenne
Rawlins
Decatur
Norton
Phillips
Smith
Jewell
Republic
Washington
Marshall
Nemaha
Brown
DP
Cloud
Atchison
Sherman
Thomas
Sheridan
Graham
Rooks
Osborne
Mitchell
Clay
Riley
Pottawatomie
Jackson
Jefferson
LV
Ottawa
WY
Wallace
Logan
Gove
Trego
Ellis
Russell
Lincoln
Dickinson
Geary
Wabaunsee
Shawnee
Douglas
Johnson
Saline
Ellsworth
Osage
Morris
Greeley
Wichita
Scott
Lane
Ness
Rush
Barton
Lyon
Franklin
Miami
Rice
McPherson
Marion
Chase
Coffey
Anderson
Linn
Pawnee
Hamilton
Kearny
Finney
Hodgeman
Stafford
Reno
Harvey
Butler
Greenwood
Woodson
Allen
Bourbon
Gray
Edwards
Ford
Sedgwick
Stanton
Grant
Haskell
Pratt
Kingman
Kiowa
Wilson
Neosho
Crawford
Elk
Meade
Clark
Comanche
Barber
Sumner
Cowley
Harper
Montgomery
Labette
Cherokee
Morton
Stevens
Seward
Chautauqua
Southwest Area
South Central Area
Southeast Area

Harvey County

The Koppes barn was built in the early 1880's for workhorses and a few milk cows. The lower part of the barn also had grain bins and a harness room. The loft stored the hay for the cattle and horses.

A number of years later, the lean-to was added for dairy cattle. The dairy was discontinued around 1970.

During the middle of the 1960's and 1970's, there were several fox families that took up residence in the unused horse stalls. They found a loose board and knew how to hide and were rarely seen, just often enough to let one know they were there.

Over the years trees have grown up around the barn since houses have taken over the farmland and it is only used for storage. The barn is still in good condition and the present owners have made a picnic area around the barn in the trees.

Miami County

A photo of this barn was sent to me to represent Miami County. I was told that this farmstead has a hundred year homestead sign on a building and that all buildings are in good condition.

I have seen a number of homesteads like this in Kansas and I know that there are, no doubt, more like this one I have yet to discover. Some of the hundred year homesteads are still in the original family and others have been bought and taken care by a new owner who has much love for the farm and it's history.

Marion County
1898
Martha Knudsen

When B.B. Neuman and Helena Voth were married on December 15, 1895, they moved onto a new farm where a small two-room house and this barn had been built for them. The barn housed his workhorses and a riding horse. The ground floor had a grain bin and harness room to go with the stalls.

The Neuman's daughter, Helen, and her husband to be, Ray Schmidt, helped her father reroof the barn. It was on that roof that they settled on their wedding date, August 14, 1948. The roof lasted until her son and husband put tin on the roof around 1990. There have been repairs to the barn over the years, but nothing has been added.

In 1977, Ray Schmidt had the horse painted on the hay door. The Schmidts lived there until 1992, when their son, Dwight, moved onto the farm. Dwight is the fourth generation to live on the farm.

The date on the picture is wrong, it should be 1895.

Harvey County

In 1869, two brothers, Sven Christopher Johnson and Andrew Johnson, arrived in the United States from Torup, Sweden. Chris was 27 years old and Andrew was 25.

They came on the train as far as Emporia, Kansas, because that was the end of the railroad. They walked to Newton and each man homesteaded 80 acres southeast of Newton in section 34, Newton Township, Harvey County. Andrew had the north 80 and Chris the south 80.

They lived the first two years in a dugout along the west fork of Jester Creek. Andrew helped Chris build a barn designed after the barns in their Swedish homeland. The barn had grain bins and horse stalls in the lower section of the south half of the barn and two rooms upstairs with a stairway leading up. One room was a tack room and the other room was later turned into a bedroom for the boys. It had plaster and lath walls and ceiling plus two windows like those in the house.

In 1873, a small home was built with two rooms and a pantry. When Chris's girls were born and the family needed more room, they added two rooms on the backside of the home. A larger home was built in 1900.

The barn and all other buildings including the first homestead are still standing and are in excellent condition. A hundred year homestead sign stands at the entrance. The farm belongs to the fifth and sixth generations now.

F.A. Wells Horse Barn

Mr. F.A. Wells came to Marion County in 1856, and bought 2,000 acres of land. A bunkhouse was built first in 1856. He lived there until his home was built. The house was built on a small hill where he could see a good bit of his land. The third story of the house was round with windows all around where he could survey what was going on. When he had his other buildings built, he watched the workers from there. If he saw someone not working, he fired that person and hired another to take his place.

The ranch had a 40' x 400' stone barn for 5,000 sheep built in the 1860's and a two story stone horse barn built into the side of a hill. The lower floor had stalls and ties for 20 horses. A well in the barn furnished water to the horses in a large wooden trough inside along one wall. Grain bins were also located there. The top floor was for Mr. Wells' buggies and for guests' buggies. The barn is still used for horses, but the top floor now holds hay instead of buggies. The long barn held pigs after the sheep and now has stalls and ties for the present owner's horses.

There was a mile of stone fence around part of the ranch. The stone fences around the buildings are being preserved, along with the buildings. The ranch still has the two barns, the bunkhouse, a smokehouse, a tenant house, Mr. Wells' home and a mill.

F.A. Wells Sheep Barn

Dickinson County

Jefferson County
Martha Knudsen

This barn is known as the Jefferson County Fair Barn.

The Jefferson Agricultural Society started the County Fair in 1864. The barn was built around 1890, and was used for exhibits from 1890 until 1908. There are stairs instead of a ladder leading to the second floor, thus, making it easy for people to go up to the second floor and visit the exhibits.

The barn was later used for cattle and horses and is still being used today. The barn originally had a cupola on it.

Leavenworth County

Martha Knudsen

A photo of this barn was sent to me to represent Miami County. I was told that this farmstead has a hundred year homestead sign on a building and that all buildings are in good condition.

I have seen a number of homesteads like this in Kansas and I know that there are, no doubt, more like this one I have yet to discover. Some of the hundred year homesteads are still in the original family and others have been bought and taken care by a new owner who has much love for the farm and it's history.

kept barn was built in the late 1840's or early 1850's as a dairy barn. was in the lower half of the barn and hay and grain was stored in the .

t owners are cleaning it up inside and may someday use it for a horse or

Wabaunsee County
S.C.P
1876

County

Sam Charlson built this barn in 1876 for horses. It is believed that the barn could have also been used as a Pony Express stop.

In later years the barn was used for cattle. The interior was then gutted and the barn is now used to store hay. Chickens were housed in the smaller building to the side of the barn.

Rice County
1900

This barn was built in 1900, for horses and dairy cattle. Grain bins were also located on the lower floor and loose hay was stored in the hayloft.

The barn has been well cared for all these years and makes a nice picture as you are driving down the highway. The owners plan to make the loft into an antique shop. They will leave the barn as is and preserve the interior structure to give the shop atmosphere.

Wabaunsee County

The home that was on this farm was built in the late 1840's or 1850's, so it is believed that the barn is around the same age. The farm belonged to a Col. Mitchell and his wife, Maude. Mrs. Mitchell was said to be an instigator in the Underground Railroad. The barn was used in this endeavor as well as the tunnel leading from the barn to the house. Original iron bars are still in place on the windows.

The barn is a historical landmark in the Wabaunsee County area.

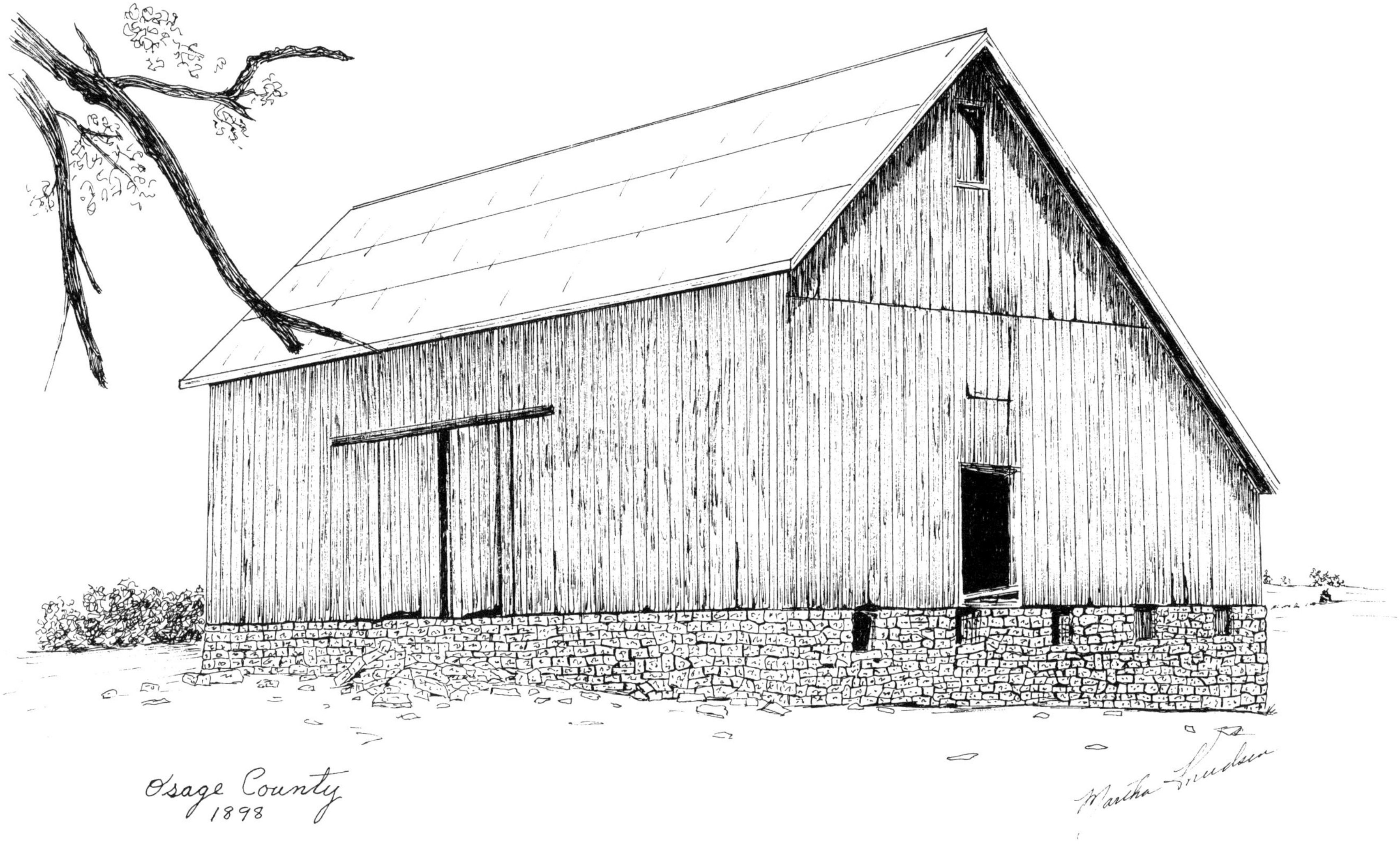
Osage County
1898

This barn near Overbrook was built in 1898. It housed horses and cattle for many years. The barn is now used for cattle, but is beginning to show it's age.

When first built, there was a ramp leading to the hayloft where wagons could be driven in to unload the hay, but sometime over the years, it was removed.

Johnson County

Most of the round barns built in Kansas were built between 1900 and 1920.

Raymond Brinkman built this Stillwell barn in 1912. Mr. Brinkman was a prominent farmer who also was a Johnson County clerk in the early 1950's. It has a limestone foundation and limestone halfway up the wall on one half of the barn and all wood on the other half. The roof is conical.

The barn was a state of the art example of agricultural technology and architecture when it was constructed. There were stalls for twelve horses, a mechanical hayfork and an elevator to hoist oats and corn to a second level for storage. Chutes from the second floor bins allowed Brinkman to distribute grain directly to the feeding stalls.

The barn took fourteen days to build and cost $1,600.

The barn is now surrounded by a new housing development. Located on the same lot as a home, it is small enough to blend in well. What it's future is, I do not know.

Harvey County

The barn was built by Bernard Stucky around 1912 and was kept in the family until the mid 1990's when his son moved into the Moundridge Memorial Home. At this time, the farm was sold.

The barn has always been kept in immaculate condition, with no weeds around it and always painted a bright red with white trim. It was built for workhorses and milk cows. In the later years it was used for beef cattle.

Geary County

William Davis Poole arrived in St. Louis in the late 1800's. He traveled up the rivers toward Kansas and ended his trip in the Manhattan area. He walked south for a number of miles to find a place to homestead. A lot of early settlers sometimes walked for days to find the right area. Mr. Poole's homestead was 2 miles from this barn. He built another house and this barn on the east end of his property to protect his land.

In 1910, Mr. Poole had this barn built to replace the one that burned in the early 1900's. It is a bank barn with a drive in hayloft on the upper level and room for horses and cows below. Over the years, a silo was added for feed. The barn and silo are all that exist now of the farmstead, but it is a landmark on Highway 177.

Graham County

Shawnee County

Martha Knudsen

Lyon County

Martha Knudsen

Charles Even, a carpenter and neighbor, built this barn in 1918. Its purpose was to house a herd of milk cows and the horses used on the farm.

The lower level of the barn is divided into three sections. The north side was for the cows, the center was used as a drive-through, and the south side was for the horses. The cattle side has eight pens, used to separate the cattle when needed. It also contains a water tank. Water is pumped in from a nearby pond. The south side has several horse stalls. Hay chutes from the loft were connected to each stall for ease of feeding.

In recent years, a change was made in the barn. After tractors replaced the horses, two stalls were changed into a grain storage bin. The center drive is now used for shedding machinery. For the last several years, part of the barn has been used for storing baled hay. The barn roof, originally shingled in wood, is now covered in tin. All these changes have made the barn a very useful building on the farm. It is very well taken care of.

Butler County
Martha Knudsen

Meade County

Martha Knudsen

Kingman County

The barn was built July 27, 1913. The lumber was shipped by rail to Zenda and Ed Tatro delivered the lumber by horse and wagon to the present sight, 1/4 mile south of Zenda.

W.G. Goenner had the barn built for a large dairy consisting of 50 head of cows. In later years, the barn was changed some for other uses.

Joe & Ida Thieme of Zenda purchased the farmstead from the Goenners.

Johnson County

Martha Knudsen

This barn is part of a late 1800's farmstead that consisted of this barn, a horse barn, chicken house, and several other outbuildings. All are still in good condition.

The barn was built for cattle with the silo enclosed. There were feed bunks on each side of the silo in the barn where the cattle were fed. The only drawback was that if too many cattle were inside the barn during the winter months, they could get too warm, causing them to get sick.

Upper Pleasant Home Barn

PLEASANT HOME FARM

This farm was originally given to a man for his services during the Rouge River War in Oregon. This man sold it to a Lucien Mansfield from whom Mr. Herman Oesterhaus purchased it in 1865.

Mr. Herman Oesterhaus came to America from Germany about 1848, settling for several years in Evansville, Indiana. As new lands opened, he pushed on to Westport, Missouri and then into several areas of Kansas before settling southwest of Junction City.

Mr. Oesterhaus was a stone mason and helped to build many of the buildings on Fort Riley, including Custer's quarters and the first Territorial Capitol. He built 2 stone barns on his farm.

His first barn was built in the 1880's high on the hill where his home was located. It was used for horses with room for hay and grain on the top floor. A small lean to was built on the one end for a milk cow or two for the family.

Mr. Oesterhaus had many acres along the Smokey Hill River. He built his own bridge across the river so they could get to Junction City and back. At that time there were many hills, as there are today, but with the many floods that came over the years, some of the hills have washed away and new ones formed. Islands were made along the low lands because of the force of the water and the construction of Highway 77.

PLEASANT HOME FARM, continued

The lower barn was built on the flat close to a sand bar where the water was shallow and cattle could reach the water easily. It was also a place where stragglers from the Santa Trail crossed. Built in1902, the barn had room for 80 head of cattle side by side in stalls. The farm, which raised registered Hereford cattle, was called "*Pleasant Home Farm*". They entered into a business with a butcher in Junction City and sold all their beef to Fort Riley for 5 cents a pound.

The lower floor was of fancy brick with a fine design and had doorways with a keystone in the center to hold the doorway upright. The upper floor was for hay and grain storage. Later a windmill on the hill was built with a storage tank underground to send water down to the lower barn. This way the water didn't freeze in the winter.

The barn was no stranger to big and small floods. The brick floor, having been cleaned many times, is now under mud since the barn is no longer used. Cattle had to be rescued at times due to the floodwaters.

As Mr. Herman Oesterhaus had 3 sons, at his death, the farm was divided into 3 parcels, one for each son. Only one son kept the land, the other sons sold their portions. The son that kept his portion had a daughter who inherited the land, who in turn gave the land to her daughter. It remains to be seen if the farm will stay with the 5th generation or end there. There are very few females today that still own the land their great or great grand parents homesteaded.

Lower Pleasant Home Barn

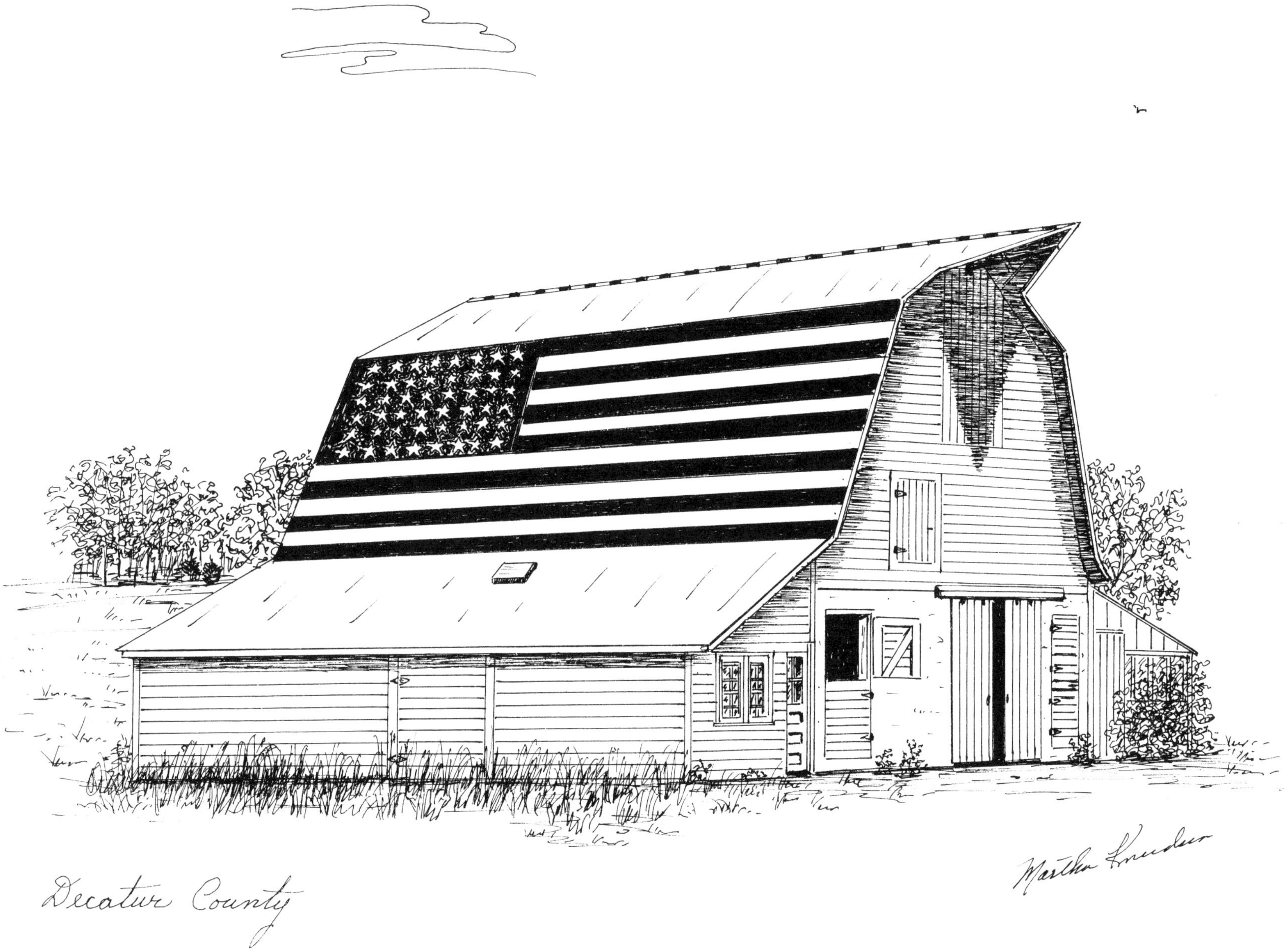
Decatur County

No one knows for sure when the barn was built, but a lot of barns of this kind were built in the early 1900's. This barn was used for work horses and milk cows plus hay and grain storage. In later years, machinery was stored in one end.

One of the early owners said she remembered many hitchhikers spending the night in the barn. Some were invited to stay and others were found there when the family went out to milk in the morning. The lady of the house usually fed them breakfast.

The last owners moved there in 1969 and also used the barn for milking, hay and grain. In early 1987 the flag was put on the west side of the roof so it could be seen from the highway. Red, white and blue colored metal was used to design the flag. The stars were cut out by hand of white metal.

What is unique about having a flag on the barn is that the family has always sold fireworks since they moved there. They are noted for living at the place with the flag on the barn. It is a nice thing to see as you drive by.

The family said they love America and it does something to them when they see the flag. They thank God for America; thus, it was fitting to put the flag on the barn when it needed a new roof.

Doniphan County

The John Fuhrken barn was built in 1915, east of Purcell. It is an octagonal structure with a circumference of 192 feet. It is one of only a few octagonal and round barns left standing in Kansas.

The barn had five double stalls and one single stall, primarily for mules. There were also two granaries and a harness room on the main floor. Two large hay doors project through the roof, so hay could be unloaded from either side.

In 1992, a Budweiser Beer commercial was filmed there. The company was attracted to Purcell because of the unique barn east of town and the rolling hills and cornfields of the county. It captured the essence of the Mid West.

D.L.SHIRK
Douglas County
Martha Knudsen

The D.L. Shirk barn was built around 1902. The farm was originally a potato farm. The stone silo like building was a water tower with a room below to wash the potatoes and anything else that needed washing.

The barn was used for horses and cattle for many years. The barn now has been remodeled inside and is used for wedding receptions, reunions, etc., but still retains its unique charm.

Harvey County

Martha Knudsen

Ford County

Finney County

Martha Knudsen

This barn was built in 1878 as the livery stable for the now Historic Windsor Hotel in Garden City. The livery stable was called "The Windsor Livery, Feed & Sale Stable" and was used until 1916.

In 1916, John Landgraf moved the barn to his ranch 16 miles north of Garden City to the SE1/4 27-21-32 of Finney County. The barn was dismantled into sections, put on sleds and pulled by horses and mules the 16 miles over snow.

The barn was used for many years, but now stands alone, to be taken apart little by little by the wind and storms. What stories it could tell if only it could talk.

Coffey County
Martha Knudsen

This barn is 100 years old, maybe more.

It was used for cows and horses. Horses were stalled on the west side and cows on the east, with hay down the center. Above the horse stalls was a storage area with a wooden floor, a ladder was used to reach this area. At the south end of the cattle area was a granary where grain was stored to feed the milk cows.

The barn was very large for the community and is the only one standing. The Floyd Calvin family bought it in 1939. They lived there for over 55 years. The family still owns the farm and the barn has been kept in good shape. It is rented out for hay storage.

Anderson Circus Barns

ANDERSON CIRCUS BARNS

As a very young boy, Fred "Bud" Anderson loved animals and would put on little circus acts for friends and neighbors. Around 1920, he started a one-man circus and by 1935, it had turned into a three-ring circus, *"The Seal Brothers"*.

Over the years he had wintered his circus in Fredonia, but wanted the winter home to be Emporia where he had grown up and where his parents still lived. He bought the former Ed Collins suburban tract east of town, and in October of 1935, started construction on the wild animal barn.

The fireproof barn had brick and tile walls, an asphalt shingle roof and steel window frames. One of the short sides of the barn faced old Highway 50. The building is located about a block from the concrete road.

The contract called for construction of two large chimneys and two ventilators on the roof. Brick was used around the openings and on the corners and the remainder of the walls were tiled with Humboldt tile. Two concrete troughs, each two feet wide, six inches deep and seventy feet long, were also built. The fireproof, heated barn cost around $2,600. The barn was completed in November 1935, and the circus moved in for the winter.

Construction of a $1,600 fireproof octagon shaped training barn at the winter headquarters was started in December 1935, and completed in November 1936. It was 40 feet in diameter and made of brick and tile.

In 1937, a horse and pony barn costing around $3,000 was built. The barn was 70 x 35 feet, almost a duplicate of the wild animal barn, with a connecting corral. The barn has nine box stalls for eighteen horses and stalls for sixteen Shetland ponies. In one corner of the barn is a two-room apartment with cooking facilities.

ANDERSON CIRCUS BARNS, continued

Bud Anderson sold his circus headquarters in 1943 to I.B. Heunvigardt of Staffordville. He then signed on with several other circuses until he sold out completely in 1945. He then bought land around Emporia and raised show horses for different stock shows. Bud Anderson was killed in Miles City, Montana in a truck accident in which seven of his twelve prized horses were also killed. He was 55 years old.

Riley County

An English Barn

When a wealthy Englishman builds a house and barn on his 4,000 acre farm, he has visions of the Baronial Estates he knew in England. Edward Davis became wealthy in America from an invention and royalties.

The 105' x 30' barn was the first structure he erected on the place in 1900. Hand carved pillars are on each side of the fourteen heavy paneled horse stalls lining the driveway through the barn. Elevator buckets take grain up to the grain bins. There are individual chutes to each of the cow stantions so the correct amount of grain could be delivered to each animal.

Living quarters were located on the hay mow floor for the barn attendants. The finest lumber was used and each nail was dipped in hot resin before being driven into a board. Only English carpenters were employed to build the barn.

Mr. Davis lived on the place for only 10 years before selling the "Ard-Mar Ranch" to Robert Mantz in 1912. A.H. Hill held title until 1918, when he sold the place to H.O. Thorne. Mr. Thorne's daughter, Mrs. Ben Freser, lived there for 40 years. Mr. & Mrs. James Peterson now own the ranch.

Atchison County
1870

This L-shaped barn, located one mile south of Larkenburg, was built in 1870, by M.E. Larkin. Mr. Larkin purchased the 130 acre farm for $4,000 from John Sheit.

The barn was built of 16 x 16 hewn white pine timbers brought in by rail from the south. Native limestone makes up the foundation.

In the early days, the barn's first floor housed purebred racehorses. The windows on that level were covered with iron bars. The second and third floors of the structure were used to store grain, hay and supplies. The barn still remains standing and in sound condition today.

Clark County

Martha Knudsen

The barn was built in 1902. The current owner's grandmother homesteaded a part of the farm.

In 1990, the farm was designated as a Kansas Tradition Farm, owned and farmed or lived on by members of the same family for 100 or more years.

The barn was built primarily for work horses, but also had space for cattle and a loft for hay.

The barn is still used and has had some repairs to the foundation, paint, and a new roof put on a few years ago.

Chase County
CROCKER
BRO'S
1906
Martha Knudsen

The Crocker Ranch was established in 1866, now only one old building remains of the original homestead.

This barn was built in 1906. The first story is made up of rubble stones and the lintels are of a heavy stone. The second floor has been constructed of board and batten that has been painted white.

The upper portion of the barn held hay and the lower floor was for work and riding horses. Today the barn still houses horses for riding.

Dickinson County

Martha Knudsen

This small stone barn was built in the 1870's for horses and cattle. In later years sheep were housed there also.

The barn had stalls and stanchions and a grain room. The loft held the hay. The north end is in good shape, but they have had to add support rods to the south end in later years.

In the early years when there was a county school a mile or so away, the kids use to stop and play at the barn on their way home. On the inside of the barn, the walls have the names of school kids and former occupants scratched into the stones.

A stone ax was used to cut the stone on the outside surface. Mr. Frye, the current owner, still has a stone axe. The hayloft is now used for his collection of butter churns as well as other antiques that he has collected over the years.

1838
Johnson County

Standing as it has for over 100 years, the Thomas Turner Barn majestically rises in a quite landscape at 199th and Moonlight Rd. It is definitely a Johnson County landmark.

A well-known area barn builder, A.J. Clinton, of Springhill, built the barn in 1898. Mr. Clinton had a trademark style he used on all of the barns he built. This trademark was a double-diamond pattern on all exterior barn doors.

This barn was built on a stone foundation with Louisiana Red Cypress. Measuring 40 by 60 feet, it was built with horse and mule stalls as well as room for loose hay. Wire hooks still hang as remnants from the days when lanterns were the source of illumination for farmers shoeing and harnessing their horses. The wooden planks inside the barn show the fine craftsmanship of its original builder. Walnut boards without a single knot outline the mangers and wooden braces support the roof.

The barn was a forerunner of modern pre-fabricated buildings. Marvin Turner, a previous owner, said he was told that parts of the barn and the cupolas were built separately on the ground, then lifted into place. Neighbors came to help with the barn raising.

The square cupolas stand ten to twelve feet above the roof. The metal roof of each cupola curves to a point, a bit like a Japanese Pagoda roof. They are each topped with a lightning rod and trimmed with shuttered windows to ventilate the huge hayloft.

Over the years, some of the stalls have been removed to make room for storage of modern machinery and feeding equipment. There are bins still used for grain storage and baled hay replaces the loose hay in the loft. Many people around the area can still remember the fun of jumping from the rafters or from the hay rope into the loose hay below.

TURNER BARN, continued

The barn cost $2,000 to build. After a tornado went though in 1981, the barn was reshingled at a cost of over $5,500.

In 1988, the Turners won a "Barn Again" contest sponsored by the National Trust & Historic Preservation and Successful Farming magazine. They were one of 85 winners out of a field of over 500 competitors.

Today, the fifth generation of Turners farm the land. W.T. Turner's son, Ollie, acquired the farm in 1905. In 1952, the farm was passed on to his son, Marvin. Marvin has passed it down to his two grandsons, David and Scott Tesenmeyer. The barn is still a show place in Johnson County.

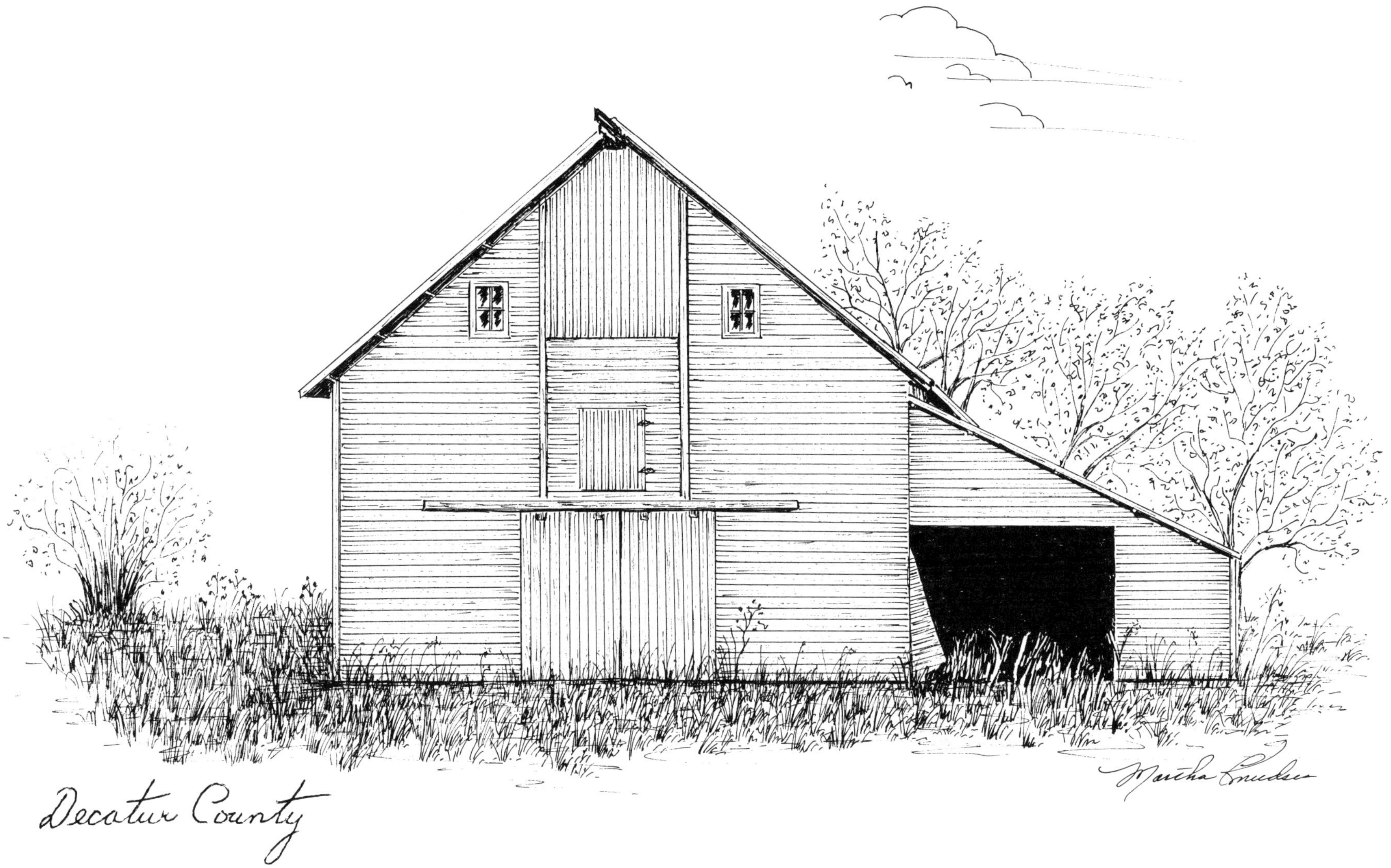
Decatur County

Riley County

Wallace County

Sedgwick County

This barn belonged to John McQuillan. In the 1920's, he was a mule trader, who bought and sold large numbers of mules.

As mules were selling well then, many farmers had their draft horse mares or other work mares bred to jack donkeys. This cross resulted in a mule. The mules were used to work on the levies on the Mississippi River.

Weanling and yearling mules were easy to bunch and feed, so Mr. McQuillan bought mules by the hundreds. It was said that his lot was so full that sometimes you couldn't count them all.

A large barn was built, as it was needed for the large number of mules Mr. McQuillan owned. The barn was divided into four areas with a long cement watering trough running through the center of each area. Hayracks were placed around the sides of each pen. There also was a large hayloft for hay.

The barn was painted and reshingled. Brush and weeds were also removed and the barnyard cleaned up. It looked like new, until a high wind came through and tore big holes in each end of the barn. The damage has been repaired and hopefully Mother Nature will be kinder to it now.

Reno County

This barn was built in 1905, for horses. The north wing contained three stalls while the south wing held four stalls. Each wing had a hayloft. The main, center portion of the barn is eight sided and open.

In 1948, a tornado took the north wing and the cupolas. Originally there were 3 cupolas, one on each wing and a large one in the middle.

The lumber to build the barn was hauled in from Nickerson and cost $1,200. The barn couldn't even be shingled for the $1,200 today.

Dickinson County

This barn was built in 1944 for Ethyl Startzman and is 35' X 69'. A wood barn of similar size and exterior design preceded it.

The stone in the barn was originally from the Woolen and Later Flour Mill in Enterprise. The mill, three stories high, was built in 1873 and razed in 1943 by Herman and Arthur Anderson. It is believed that these two gentlemen built the barn.

Ethyl Startzman was a progressive farm lady, doing much farm management personally. She was know in the area for having a quality Hereford cattle herd. She kept a few dairy cows, mostly for personal use and some chickens also. She was known as a hard worker, shrewd in dealing, but also very kind and generous. She has also left her legacy in a college scholarship given each year through Dickinson County 4-H.

The barn features 3 front stalls with large hay bunks and sliding wood doors opening into the main hall. The Mel Purdue family, who now owns the barn, added steel gates and 2 stalls, creating a new hallway. These stalls are used for horses.

The barn still has 3 original milking stalls, which have been used for rabbits. The main floor is all concrete. Other features on the ground floor are 2 grain bins and a sliding door toward the back. Feed bunks at the back are open to the loft for a hay drop. In 2 places near the front stalls, the walls are only 1/2 as thick as the rest, creating a shelf for tools or medicine.

The loft is full size except for one grain bin in a corner. This bin has an exterior grain spout built in and an opening for an auger to put grain into it. The loft has a single bale drop hole in front of the front stalls. The Purdues use the loft for hay and straw, but also for 4-H meetings, dances, basketball, pool, dummy roping practice, exercise equipment and storage.

The Purdues bought the barn in 1990 from Clyde Haynes. Mr. Haynes had completed an extensive cleaning and basic maintenance on the barn with the expectation of converting the barn into a business. The Purdues had two boys with several 4-H animals, so they were grateful to find this barn for sale. The barn is well cared for.

Meade County

Built in the early 1900's, this barn was for horses, cattle and hay. In later years, a ticket booth was put at the top of the loft stairs and barn dances were held up in the loft. The ticket booth is still there today, no doubt waiting for another barn dance to be held.

If barns could talk, they might have a lot of happy and funny stories to tell.

Harper County

Mr. C.C. Phelps completed the largest livery and feed stable in Harper County in downtown Harper on November 7, 1878. The building is 100 ' 5 1/2" x 49' 6", strongly built and of good material. The best grade of shingles was put on the roof and some of the best carpenters exhibited their workmanship in its structure.

The red barn served early pioneers arriving by train as their place to purchase horses and wagons to further their journey to settle this area of the state and in northern Oklahoma. It also housed a horse drawn "bus" service that made six runs daily to the train station to pick up travelers and take them uptown.

Around 1928, after the arrival of the "horseless carriage", the barn was purchased at a Sheriff's sale by a local businessman and given to the City of Harper. Shortly thereafter, it became the exhibition building for the Harper County Fair.

In 1990, the barn was turned back over to the city as the Harper County Fair Association decided the Red Barn no longer fulfilled their needs as an exhibition building and updating it would cost too much.

Friends of the Red Barn continue their efforts to raise funds to preserve and maintain the Historic Red Barn by holding all kinds of fundraising events there.

Lincoln County

Martha Knudsen

The father and uncle of the present owner built this barn in 1937. The stone came from the quarry of the owner's grandfather, 1 mile west of the barn.

When laying up the walls, 12-inch stones were used. This is about half of the thickness of the walls in other stone barns. The dimensional lumber came from an old livery barn in Lucus. People told him it wouldn't stand up, that the walls were not thick enough. Today, the barn is still very sound and only one or two small hair line cracks exist. The original barn roof has been replaced with steel.

Stafford County

Martha Knudsen

This barn was built in 1910, and was divided in half, one half for cattle, and the other for horses. It is 50 feet in diameter and 42 feet tall. A silo was built in the center of the barn and hay was stored around it on the top floor.

The loft was filled with hay from inside the barn.

Years later, I'm sure most of the interior area was used for cattle. The barn is not in use now, as like so many barns of today, it has outlived its usefulness. It is falling into disrepair.

Butler County

Martha Knudsen

WHITEWATER FALLS STOCK FARM

J.W. Robinson emigrated from Aberdeen, Scotland. Somewhere in the immigration process, his name was changed to Robison. He first settled in Illinois where he became a senior partner in a law firm.

In 1878, he came to the future town of Towanda, Kansas and bought land along the Whitewater River. He called his farm, "Whitewater Falls Stock Farm". J.W. couldn't resist the acres of alfalfa, bluegrass and hilly slopes that contained an abundance of limestone.

J.W. bought farmland for $1.50 an acre from Charles Potwin, founder of the town of Potwin. At first he planted an abundance of wheat, causing the town folks to nickname him "Wheat Robison". He also had a good supply of cattle and other livestock. He had a good personality and reputation that eventually won him a seat in the State Senate.

The farm, Whitewater Falls Stock Farm, was named for the river that ran through the farm and the three-foot high falls naturally flowing from the river onto the farm. He at first hired 2 men to help him farm and during the productive years, employed 4 or 5 men and their families to work on the farm. These men lived on the north and east ends of the farm. During the early years, J.W. made an important purchase of Percheron horses. This venture was such a success, that he soon went into the business of buying, selling, trading and showing horses, mainly Percheron horses.

Soon after he got married, J.W.'s son, J.C., helped run the farm in partnership with his father. J.C. & J.W. often traveled to Belgium and France to obtain the type of horses they wanted. When the Robison's were well into the horse business, there were often two to three hundred horses present on the farm at all times. The W.F. Farm was known to have the largest purebred Percheron herd in the west, and at one time, the most prestigious in the USA.

WHITEWATER FALLS STOCK FARM, continued

One of the farm's greatest purchases was a horse named "Caseno". Caseno was bought in 1903 and won worldwide fame. After winning many top ranking shows in America and France, including the First World's Fair in 1904 in St. Louis, J.C. was offered $10,000 for Caseno, but he quickly turned down the offer and went on to win more shows with Caseno. There were also other winners on the farm, stallions, mares and even colts. Junior Robison, J.C.'s son was the main man when it came to taking horses to these shows and fairs.

W.F. Farm had a variety of horses and livestock, such as Clydesdale horses, 2,000 head of cattle and 1,000 head of hogs. Buyers came from all over American and even the world to purchase the Robison's top quality stock. The sales pavilion was 40 x 100 feet.

In 1909, J.W. died and the farm was left to his son, J.C. The rest of the farms in the surrounding area that J.W. owned were divided up among the rest of the Robison children.

Social events were always occurring on the farm and it was the center of community entertainment. In May 1934, J.C. and Will Pierce organized the W.F. baseball club called the Robison Percherons. Every 4th of July, the farm hosted three to four thousand people for a big picnic. People sometimes came from different states and countries. Loft dancing was held in the barn. One of the well-known events on the farm was the auctions that took place there. The farm would be decorated as if a circus had come to town. On March 1934, the greatest sale ever of Percheron horses took place on the farm with some 48 horses being brought in.

WHITEWATER FALLS STOCK FARM, continued

During the depression years, the farm still kept going and when someone ran low on feed, they came and asked J.C. for help. When farming became more mechanized with gas powered machinery, horses were no longer used and traditional ways of farming became obsolete.

The decline of the W.F. Farm did not come as a shock to the Butler County community. After J.C. took his life in 1937, because of health and financial difficulties, the farm was left to his five children. Only one, Junior Robison, was able to take it on. He sold the horses and raised Hereford cattle. After ten to fifteen years, times were still changing rapidly and the farm was difficult to keep up. Junior sold the farm to Peter Zuercher in 1945. Zuercher did not keep the farm running and ended up, years later, selling it to the Foley Tractor Company. The Fagen family leases the farm from Foley Tractor and are the present residents.

Doniphan County
1871

This mill barn was constructed in 1871, as a flourmill on Rock Creek with a mill dam on the 64 acres purchased by Abram Detwiler of Philadelphia in 1868.

The building was turned into a barn when milling was discontinued. It is unknown when this conversion took place. It now sits idle, but is in remarkably good shape considering its age.

Reno County

Built in 1912, this barn was used originally for dairy cattle and maybe a few horses for work, riding or both. Loose hay was put into the loft for feed.

The barn is still in use, but for beef cattle and calves. The hayloft now holds small square bales for feed.

One year a tornado pushed a corner of the barn off the foundation, but they were able to place it back on the foundation and support it with a steel beam

The barn is kept in good condition by the fifth generation and their children. These children, the sixth generation, have spent many happy hours in the barn playing with kittens and feeding calves.

SINCE 1903
MILLERS
OF CLAFLIN
FURNITURE CARPET DRAPERIES
Barton County

Mr. Bobby Joe Miller had grown up seeing Red Man Tobacco signs and other advertisements painted on barns. Thus, he thought it would be an appropriate way to advertise *Millers of Claflin*, a business dating back four generations to 1903.

Mr. Miller's sons, Brad and Bill, said it was their father's idea, something he always wanted to do. The sign was completed in 1987, and was the last thing Mr. Miller did before dying of cancer. While fighting the cancer, Mr. Miller would drive by the barn, as it would always put a smile on his face.

In 1994, a field agent noticed the sign on Highway 4, and said it was estimated only 128 feet from the road. This was in direct violation of a 1972 Kansas beautification law restricting highway advertising. LadyBird Johnson had originally put together the beautification program. Mr. Miller had written the State of Kansas before the sign was painted, asking if there were any prohibitions against such signs and never received a reply.

Brad and Bill Miller had to go though many people and a lot of red tape to keep their sign. They were able to accomplish this by adding a line at the bottom of the roof stating *"A William Miller Memorial"*.

Harper County

Martha Knudsen

KELLER BARN

The Keller barn, as it is known, was built in 1908. It is 72' in diameter and 52' high. Originally built for dairying, 70 head of cows could be milked at one time. Horses were also kept in the barn. A manure gutter and trolley system along the outer wall inside the barn made removal of the manure easier.

The barn was well built and has been well taken care of. When you step inside and see the rafters in the self-supporting roof, you understand why it has withstood all the winds and rain of time. The roof was reshingled in 1950, with 82 squares of cedar shingles. 350 tons of baled hay can be stored in the loft.

Lew Babcock, who sold the barn to Oliver and Paul F. Keller in 1924, built the barn. This is the smallest of the 3 round barns built in the Harper area.

Mr. Babcock came to Harper in 1879, with his father and 4 older brothers to buy cheap land. Babcock got a job as a cook for the Sons of British Aristocracy. This organization had established a small colony at Runnymede, several miles northeast of Harper. When Mr. Babcock started farming, he traded his horse for a team of oxen and his gold watch for seed and a plow.

Neosho County

Not much history is known about this barn. The place was homesteaded in 1860, so the barn likely was built sometime in the 1860's. It was mortared and pegged (no nails were used in the framework).

The present owners bought the farm in 1965 and are the fourth owners. Hay and grain were stored above and a Grade A dairy was run down below.

As with a lot of haylofts, after the hay was gone, the kids played basketball. Many a game was played there over the years.

Lane County

Martha Knudsen

Joshua Wheatcroft built this barn around 1919, for cattle and horses.

The barn is in good shape and people still remember the dances held in the barn loft. Food would be brought and put into the house until the dance was over and people were ready to eat. Many times the dances lasted until dawn and good times were had by all.

On one of the pillars in the barn is a sign *"Please Do Not Spit On the Floor"* signed *"The Management"*. They say the sign is still legible today.

Douglas County

Eben Baldwin built this barn in 1879. The inside structure of the barn has original hand hewn green trees for rafters and supports. It appears the trees were oak, hickory and walnut. Some still have the bark on them.

The barn is built into a bank and on the west side it opens into a full basement with steps leading to the main floor. There are three doors into the basement on the west. It is said there were once stalls for 30 horses in there.

The main floor has two small rooms, one on the south side, then an entryway on the east and a corncrib on the north, plus a stairway leading up to the third floor. There are 2 large doors in the middle of the north and south sides. The hayloft takes up 1/2 of the barn's second and third floors.

The east door opened inwards and was large enough to drive a hay wagon or buggy through to the north door. In later years trucks loaded with hay could drive through the north door to unload baled hay, and then back out. The hay track and fork are still in the top of the barn. It could hold up to 80 tons of loose hay.

Many people recall going to parties there in the earlier days. Once in the country by itself, it seems that houses are slowly surrounding the farm and barn.

Morton County

Martha Knudsen

This barn was built in 1920, by a carpenter who lived a mile east and walked to and from work every day. The trench for the foundation was hand dug and the sand for the cement was hauled in from the Cimmaron River, six miles north, with a team and wagon. The cement was all hand mixed in a mortar box with a hoe and shovel.

The barn was used to shelter a lot of work horses and store grain and hay until the mid 1930's. In later years, more grain storage was added in the barn.

The barn was also used as a landmark in giving directions to the owner's home and to other's homes as well.

RIDDER
CATTLE
Ridder Cattle
Wichita County

This barn was built in the early 1920's on the Hackmaster Ranch. Mr. Hackmaster was a wealthy man from Florida who owned race horses. The barn was possibly used for his racehorses and for his registered Hereford cattle.

In the late 1930's or early 1940's, the barn was moved several miles south and a little east to its present location.

The Ridder family purchased the farm and barn in 1956. They have used it for livestock of all kinds. When they first moved there, they used the tack room as a brooder for baby chickens for many years. When the chickens were large enough, the roosters were butchered for food and the pullets moved to the chicken house to lay eggs. The horse stalls were replaced with stantions for milk cows. Cream and eggs were sold to buy groceries and was usually the main source of income until the crops were harvested.

In March of 1957, there was a big blizzard and all livestock was put into the barn to save them. Snow buried the hog house, so the hogs were dug out and moved to the barn across huge snowdrifts after the blizzard. The chores were all done under one roof for a time.

The children and their friends, like all farm kids, enjoyed playing in the barn. The barn still has hay stored in it for the horses and other animals on the farm in case of a bad snowstorm.

The barn has been rebraced, steel siding added, and a tin roof put on to preserve it for hopefully a long, long time to come.

Mr. Ridder loved his barn. As a tribute to him, the family had the barn's picture reproduced on his tombstone.

Cloud County

Charles Lagasse Sr. and wife, Leonie, immigrated to America from France in 1870, when Charles Jr. was a small child. They homesteaded on a quarter section of land in Cloud County. They lived in a dugout for a time until they were able to build the frame house that still stands on the place.

The front view of the barn in the picture is the oldest part of the barn and has the date, 1897, painted on a beam. This side of the barn has four double tie stalls that would hold eight horses. The other side held stalls for four horses. The second floor above was large and open to store loose hay. It had holes in the floor for hay to be dropped to the stalls below. In later years, hay bales were stored up there.

The milking parlor next to the older section of the barn has wooden stanchions to hold seven cows. It has a cement floor with a channel that ran behind the cows as they stood in the stanchions. The channel drained to the outside so the floor could be kept clean easily.

Another addition was above the separator room next to the milking parlor where a loft was added to house a large metal water tank. The windmill pumped water up to this tank. The gravity flow from the elevated tank provided running water to the house, milking parlor and stock tank. It also may have contributed to keeping the separator room cool. The separator room was a small room that housed the separator. When the milking was done, the milk was poured from buckets into the separator where the cream was separated out. At first this was all done by hand crank and later electricity took over and also added lights to the barn.

The east side of the barn has a loafing shed. An addition at some later date put a haymow over this part. There are mangers along one side of the shed and the mow floor above has openings to drop hay into the mangers below. Further additions were added to the original barn over the years and in 1910, the large barn on the backside was added.

The barn is now in use for hay storage and horses. The milking parlor may become a chicken house.

Atchison County

Martha Knudsen

Marion County
Martha Lueders

Edwards County

Martha Knudsen

Greenwood County
Martha Knudsen

Pawnee County

Martha Knudsen

Wallace County

Scott County

Martha Knudsen

Comanche County

Wabaunsee County
1905

This barn is being restored.

Peter, J.L. & Friederick Muehlenbachor (all Civil War veterans), along with a sister, Margaretha, homsteaded along a creek about seven miles southwest of Alma.

Margaretha was allowed to marry a Mr. Schepp, as none of her brothers had married and they wanted an heir. Her son, John Ludwig (Louie), was born in 1867. When Louie was small, the brothers chased Schepp off. Schepp went over to Deep Creek and married again. (Louie had two half brothers).

Louie inherited the ranch from his mother nad it was known as the *Schepp Ranch.* It had many pastures and was ideal for steers shipped in from Texas.

The three story barn was built in 1905. A driveway to the third story allowed hay from a wagon to be tossed into the second story hay loft. In later years, a hay fork was added to place the hay in the loft. On both sides of the driveway, there were grain bins for storing grain for the horses and cattle.

1886
McPherson County
Martha Knudsen

Not much could be found out about this barn, only that it stood in Moundridge. It may have been built where it stood. At one time, a Dirks family lived there.

The barn was moved to a farm south of Galva at some point. Once there, it was restored on its original foundation and painted. This is a small barn that was very useful in its day as well as today.

Haskell County
Martha Knudsen

Greeley County

Martha Knudsen

Leavenworth County

The barn was built after World War II in the late 1940's or early 1950s. It was built as a dairy barn and was used as a dairy for many years. The inside still remains the same, but it hasn't been used for a number of years. It is well taken care of.

Chase County

Oscar Duehn was born in Germany in 1856 and came to this country in 1863. He homesteaded in Chase County, probably in the 1870s, as he lived on the farm where the barn is located for some 60 years.

The barn was built in 1913. Wood was brought into Clements by railroad where Mr. Duehn unloaded it from the cars and took it to his farm.

The barn is a three-story bank barn. The lower floor has stalls for the draft horses and well as saddle horses. An overhead rail and bucket system was used to take the manure outside, making it much easier to clean. The second floor has a drive through with two large grain bins on either side. The third floor is the hayloft.

Mr. Duehn died in 1938, and his wife in 1946. Their four children, Frank, Nellie, Howard, and Mable, none of whom ever married, continued to live on the farm until their deaths. They farmed with horses, never using tractors.

The barn is in good condition and is being restored to its original beauty by its present owners.

Seward County
Martha Knudsen

This barn is a "new" barn considering the ages of the majority of barns in Kansas. It was built in 1950, probably to take the place of an older barn. The barn has been converted to modern day use and is kept in good condition.

Sherman County
Martha Knudsen

Clay County
Martha Knudson

Nemaha County
Martha Knudsen

McPherson County

Gray County

Logan County

Franklin County

Martha Knudsen

Washington County
Martha Knudsen

Hamilton County

When Lake McKinney was being built in 1940, Mr. Hartshorn got wood from buildings being torn down to make way for the lake to build this small barn

The barn is his pride and joy and is well taken care of.

Stanton County

This barn on an abandoned farmstead looks to have been redone on the inside for a workshop and smaller equipment.

It is nice that they have saved some of the larger barns to be used like this. In a number of areas around the state, the small barns are of no use for large-scale farming, so they are torn down or left to fall down.

Barton County
Martha Knudsen

The gentleman, who built this barn in the early 1900s, built his barn to last through the many generations.

The barn was built along the Walnut Creek with a cement lower floor to prevent floods that were sure to come in the years ahead from washing away the barn and for easier cleanup afterwards. The house sits on higher ground.

The barn has a new tin roof and is well taken care of and is still in use.

Linn County

Martha Knudsen
Marion County

Smith County

Martha Knudsen

Jackson County

Sedgwick County
Martha Knudsen

This barn was built in 1915; an older barn still exists that was built around 1904-08 for horses. The two barns are in good condition and the farmstead well maintained. The farm is still in the same family.

The 1915 barn was built for a dairy. The dairy, called "The Highland Dairy", was in existence until 1929. The dairy served the Wichita area. The farm is now on the edge of one of the small towns north of Wichita.

After 1929, the barn was used for family milk cows and now, years later, is used only for storage.

Brown County

Martha Knudsen

It is not known exactly when this barn was built. The Victorian home next to it was built in 1915, and the barn was built well before the house, which was built to replace the old home that was originally there.

The barn is smaller than many of that time, and rather plain. It has board and batten siding. There are three rooms downstairs with stairs leading to the loft. There is an opening to the loft along the south wall, above a rather unusual door, which rolls on a tract around to one side of the room.

The barn, as most, needs repair and the owners hope to restore it in the future.

Elk County
Martha Knudsen

Not too much is known about this barn, however, there is no doubt that it was built for horses and cattle. There is a stone that has the date, 1924, carved into one of the walls, but whether that was a birthday or a date to remember remains a mystery. It seems unlikely that it was the year the barn was built.

The barn was also used to house goats in the later years. The present owners have cleaned the barn and fixed it up for their horses. They say it is still in sound condition and they added a new roof.

Allen County

Sumner County

Russell County

Norton County

Martha Knudsen

Cherokee County

This barn in Cherokee has had galvanized steel put on the sides to preserve the barn. It is a red barn and from a distance you wouldn't know it wasn't wood. The roof has also been tinned.

Wyandotte County

This white barn, built in the 1920's for horses, sits in the midst of a housing addition now. Its fate is unknown; however, it is in good condition and could be converted into another use.

Stevens County

This barn is located at the edge of a small town, and appears to have been modified for use as a garage and workshop. The building may be used for the towns building and road equipment or for private use.

HALEY'S RANCH
Sedgwick County
Martha Knudsen

Haley's Stable was built in 1928. The owner had great hopes for a wonderful racehorse, breeding and training operation.

The size of the structure is an imposing 120 x 30 feet and 45 feet high. Sixteen full box stalls held the horses. There were grain bins, a harness room and an office on the ground floor. Changing rooms for the jockeys were above along with the hayloft. A 1/2-mile track was located to the rear of the stable. The earth for the track's surface was hauled in from Missouri to make the footing better.

The Olympic Horse Squad trained here and the Masonic horse group stayed here when on tour.

It was reported that one famous horse was buried standing up somewhere on the premises.

After the racetrack was closed, the owners still had horses for some years, but now the barn isn't used for much as they are now raising cattle. The barn is all that is left of the racing operation.

The barn is in need of repair as wind has damaged the roof at the back and a number of windows are boarded over and others need fixing. It is unknown if restoration is in the future for this barn.

Anderson County

Martha Knudsen

Crawford County

Martha Knudsen

Ness County

Rush County
Martha Knudsen

Barber County

Fire destroyed the first barn on the farm and when they rebuilt, they wanted something more fire proof.

This barn was built in 1934, by the owner's grandfather. Red tile was used for the lower floor and tin for the loft and roof. The lower floor has room for horses and cattle, plus a grain bin. Hay was stored in the loft above.

Jefferson County

Will Parker built this barn in the late 1880s. It is 40 x 60 feet. The loft was used for hay storage, the main floor for horses, and the lower level for cattle and sometimes hogs. Will's son, Don Parker, took over the farm in 1935. As tractors became more common, most of the structure was used for cattle and hogs.

The Kansas 4-H Foundation purchased the property in 1972, for use as a wilderness camp for boys and girls. Concrete floors were poured throughout the structure and a kitchen and bathrooms installed. The barn was then used for counselor housing and storage of equipment.

In 1986, the present owner converted part of the barn into a woodworking shop for his hobby and the rest is used for equipment and lumber storage.

SINCE 1859
PIONEER BLUFFS
SINCE 1859
1872
Chase County
Martha Knudsen

Charles Rogler settled on 140 acres of land in 1859, which in time, grew to 2,720 acres. Mr. Rogler lived in a small cabin until he built a house and barn in 1872. There was also a stone wall along the trail.

The old barn (inset picture) was one story with a gable roof and sat on a grout foundation. A pent roof lean-to projects from the building's center aisle, at full wall height. It had a corncrib with a loft above. The floor was open in the center bay of the building. Walnut pegged hewn posts and beams were incorporated into the barn's construction. A tin roof was later added. The barn was moved in 1915 to accommodate the construction of a main barn and it was relegated to secondary stalls and storage status.

This new barn was built in 1915. It is a two-story gambrel roof barn that sits on a limestone block foundation. It has a western façade orientation and is 70 feet long and 40 feet wide. The barn was built for general livestock shelter and hay storage. There is a concrete floor in the traffic areas and dirt floors in the livestock areas. The first floor also has space for implement storage, a granary, a milking area, horse pens, and a harness area.

The barn originally was painted gray with white trim and wood shingles. The barn today, has a tin roof and is white with black trim.

1886
Rice County

This is another example of an 1800's barn. Some were quite large and other small, depending on the needs of the owner.

In this area, wood was more plentiful than stone. Tree trunks were often used as supports. Cottonwood, walnut, oak and pine were used in the barns.

This barn was torn down in the middle 1990's when its use was no longer required.

Butler County

This roundtop barn was located two and a half miles east and one half-mile north of Whitewater. It was built for Abraham H. Klaassen after a devastating tornado on June 8, 1941, destroyed the existing barn and almost all other buildings on the farm, as well as fourteen other farms in the area.

Abe Klaassen and John Thiessen, a neighbor who also lost his barn, went to Wichita to see the laminated building rafters that the Rock Island Lumber Company had developed by taking 6 or 8 pieces of one by two inch boards glued together. The lumber company cut angles in the arch to give it the desired height. Rock Island also sold used lumber from old railroad boxcars. The entire barn was constructed from this used lumber. The joists were 4 x 12-inch timbers from the bottom of the railroad cars.

The carpenters were local Elbing men, Elmer (Butch) Lambert and his sons, John and Hubert. Abe Klaassen's sons, Irvin, Harry, Herman, Roland and Ted also helped in rebuilding the barn and other needed buildings. The family first built a one car garage for a kitchen and living arrangement for the parents, the boys slept in a partially destroyed granary that was still standing.

The barn was 40 x 64 and loose hay was pulled into the hayloft by a grapple fork or sling. A 100-mile per hour wind destroyed the barn in 1990. This time, it was not rebuilt.

Ottawa County

Phillips County
Martha Knudsen

Sheridan County
Martha Knudsen

Logan County

Rawlins County

This basement barn was built in 1916. The barn was used for cattle, workhorses and several riding horses.

In the early 1930s, a tornado took off the top part of the barn. It was rebuilt to its present construction.

In later years, several milk cows also used the barn. The milk was used for the family while they were raising their five girls.

The girls had a sleepover for some FHA girls in the loft and fun was had by all. The girls probably played in the loose hay by jumping or swinging from the hay rope. Basketball was also a fun game played in the loft. Loose hay in the barn has given way to square bales.

The present owner was born on the farm and moved back after World War II in 1946. The buildings and farm are well taken care of.

Marshall County

This barn is reportedly the world's largest round barn. It was built in 1909, near the old town of Irving. The barn measures 100 feet in diameter and rises 80 feet into the air. The barn was built by Fred Cottrell, a breeder of registered Hereford cattle, as an oddity to attract people to his livestock sales. Mr. Cottrell was known around the Blue Rapids area as a "good operator".

A barn warming was held at Fred and Kate Cottrell's barn on August 3, 1909. Nearly 500 people reportedly attended the dance. The barn was talked about as the "Landmark of the Century". It was featured in Ripleys "Believe it or not" in 1937.

Built at a cost of $5,000, the barn sat on a locally quarried rock basement with hay and grain storage on the ground level. 500 tons of hay and 10,000 bushels of grain could be stored in the barn. By having the hay storage at ground level, it was quite efficient to drive the horse drawn wagons into the barn to unload the loose hay. Due to the weight of compressed hay today, the floor would probably have been too weak for baled hay. The large domed roof is supported by a series of truss rafters providing a clear span across the barn. In modern times, this size of clear spans in roofs is common, but at the turn of the century, they were unusual and seldom done.

The ventilating cupola forms a ring, which ties all the rafters together. Originally the barn roof was covered with galvanized iron. This leaked, so cedar shingles were then installed. From the eaves down, the barn was sheathed with corrugated galvanized iron sheets. The lower level, or rock basement as some called it, had thirty windows. Box stalls lined the outer walls and inward from there was a nine-foot feeding alley and then, a second ring of stalls. The center section housed a workroom. The barn could hold three hundred head of cattle.

COTTRELL BARN, continued

The Cottrell ranch was known by that name locally even though it was sold in 1954 to the Wacek family. The farmstead on which it stood is now abandoned and there are no longer roads to it, only a trail through a great expanse of fenced pasture on which trespassers are not welcome.

The Cottrell barn had not been seen at all by the general public for 35 to 40 years and even the people of Marshall County were fast forgetting the once famous landmark. What an ironic fate for a barn that was built to attract as many people as possible. The barn was torn down in 1995.

Republic County

Pottawatomie County

John McKimens built this barn around 1858. He was originally from Westmoreland County, Pennsylvania. Mr. McKimens named the town, Westmoreland, for his home county.

The barn is a forebay barn built after the Pennsylvania barns, but the side shown in the picture does not show that. Many of the beams are oak and some of the floor joists for the second floor are log. Some of the logs still have the bark on them. The second floor has grain bins and devices to carry the grain to the lower floor. All appear to be the original workings. Hay was also stored above.

A single story, four-square house was also built along with the barn. The home also served as the first post office of Westmoreland. In later years, a road was built between the house and barn, as a road to town was needed.

Rock Creek ran through the property and Rock Creek Ford was a campground for travelers on the "Oregon Trail", also known as the old "California Road". The trail ran at the base of a hill now called Mt. Ephram and through Pottawatomie County. A suitable camping ground was offered to freighters for their teams on the north side of the creek as the ford was a rock bottom one and an abundance of good water for all purposes was always available. Stock could also always find an abundance of grass close at hand. At times, the ground was literally covered with big prairie schooners and freight wagons. From the ford, the road then swerved to the west onto what is now the southwest corner of Westmoreland and on towards the Blue River.

The McKimens had a son, but he didn't farm as industriously as his father and soon moved on to the Oregon, Washington area. The farm was sold around 1921-22.

A few floods have changed the level of silt between the barn and creek. However, the area is still beautiful with the creek, walnut trees and old barn causing one to sit back, relax and reflect upon the past. The new owners purchased the farm in 1997.

Prospect Farm

1878

1882

PROSPECT FARM

The Prospect Farm barns were in my first barn book, but I obtained some family history on how the land was purchased and how Prospect Farm came into being. I thought everyone might enjoy reading about it. There are five or six original buildings remaining. Shelly Taylor, Jackie's great, great grand daughter, sent me the information.

Jackie Taylor and the majority of his family came to the United States in 1843 and settled in Cincinnati, Ohio. In the 1870's, Jackie believed the next area of growth was in the west. Railroads at that time were coast to coast. In the spring of 1871, Jackie and his brother, Joseph, came upon an advertisement of the Missouri, Kansas, and Texas (MKT) Railroad describing the magnificence of the land along its tracks. "MKT offers 1,300 acres of the best land in Kansas, where winters were short, dry, and pleasant, snow usually lying on the ground but a few hours at the longest." Where "Osage Orange can be set out for $0.35 a rod and in four years will turn stock". "The soil is deep and fertile and rainfall is fully as plentiful as in Ohio".

The railroads provided free transportation, or at least only a modest cost, to bring the prospective buyers on excursions to show them land and convince them of its investment value. Jackie, with his oldest son, John Hiram, and son, Joseph, first arrived in Abilene along with others comparing this land, county and state to the one they had just left. Located in Junction City was the office of Capt. A.C. Pierce, who was the realtor in charge of sales for the MKT land. They rode for two days looking at the prairie lands and grass that grew belly high on a horse. They found the land to their liking and negotiated a deal and in March 1872, Jackie bought Section 6, Union Township, Dickinson County for $3,176.20 and Joseph bought the section north for $2,656.23. Later in 1872, Jackie returned to Kansas and bought a total of 3,042 acres for a total price of $11,110.77 (about $3.65 per acre). This included various parcels of land stretching nine miles north and south.

PROSPECT FARM, continued

In the spring of 1872, the first prairie land had been plowed, 100 acres, using a walking plow. Spring wheat was then sowed. Later, hard winter wheat was found to be better. Jackie went back and forth to the farm from Cincinnati until 1885, when he settled in Kansas.

On September 28, 1877, Jackie Taylor deeded all of Section 33 to his son, John Hiram, who was 20 years old. His first project was to drill a good well, 50 feet southwest of the house. They built a section of the house in 1876 of native stone from Swenson Creek, four miles west. It was used as a home for eight years. The cellar, as a kitchen, had the bake oven built into the outside stairway. The first floor was used for living quarters and the second floor, just one large room with beds, was used as a bunkhouse for fired hands. It took many men to build a farm and to till the soil with horses.

Construction of the first barn was begun in 1878 and finished in 1879. The barn is 50 ' x 200' feet with upper floor for grain, hay and equipment. The lower floor could house as many as 50 head of horses and mules needed at plowing time and as many as five teams of three horse hitches were used at any time.

The second, or east barn, was finished in 1882. The two barns were somewhat similar as they both had a driveway to reach the upper floor, with cisterns for collecting rainwater located under the driveway. This barn was used for Shorthorn cattle, hogs and sheep. In later years, a silo was added.

John Hiram's vision for Prospect Farm was to supply his family with wholesome food. He planted orchards of apple, pear, peach, plum and cherry trees and a large vegetable garden. Each summer, hundreds of quarts of vegetables and fruit were canned for use in the winter.

PROSPECT FARM, continued

John Hiram married on August 20, 1884, in Salina to Fannie Converse. Shortly after the marriage, he built a beautiful stone house from the stone quarry eight miles northeast of the house. It is shaped like a "T" with the new part of the home attached to the original house built in 1876. The front of the home is 40' x 40', three stories tall with a cupola at the top. There is a huge basement under the entire home, the builders had to dig 11 feet deep, going through three solid rock layers, each 12 to 15 inches thick, to reach bedrock. The solid foundation has kept the home from showing any cracks for over 90 years and up to today, I'm sure. The cream colored limestone was dressed with a five tooth chisel, burbs about an eighth of an inch long with trim all around the edge. Each of the fourteen rooms had several windows, each window has a stone lentil, hand carved in the designs of the day which portrayed signs of prosperity, health and happiness. The lentils show the ivy, the thistle, the heather, an anchor, a star, etc., each having its own significance. Over the front door was a large stone carving with the initials of "J.H.T." One story about the carving of the lentils says that John Hiram was getting anxious to have the house finished and asked the masons to duplicate the designs and they said "no, it would be bad luck for you are going to have ten children", which is just the way it happened. The stone masons were paid $4.00 per hour.

The floor plans were on simple lines. The first floor had two large rooms on each side of a spacious hallway. Three rooms were finished in black walnut and the dining room was wainscoted in cherry. There are four square rooms on the second floor with a wide hallway and stairway; all are in yellow pine stained with black in the grooved woodwork. The third floor, also in yellow pine, was composed of three bedrooms with dormer windows. A small one-room cupola was built on top of the house. It had windows facing each direction so one could see the surrounding prairie for miles.

PROSPECT FARM, continued

The home where John Hiram grew up had a fireplace in each room. In contrast, this home had no fireplaces, but was heated by coal burning baseburners`, two stoves and the kitchen range. A dumb waiter was built to carry coal to the upper floors.

The cash crop was hauled by wagons to Skeddy, sometimes twelve wagons left at a time. They were loaded on a railroad car and picked up by the MKT Railroad. In the late 1880's, a Rock Island branch line was built at a little town called Pearl, named after Jackie's daughter. The little town was closer to the farm than Skeddy.

In 1963, Mrs. John Taylor celebrated her 100th birthday, having lived on the farm for 79 years. The last Taylor living on the farm was Paul, John Hiram's son. Prospect farm was in the Taylor family from 1872 until Paul died in 1995.

Labette County

Martha Knudsen

Barton County
Martha Lunsden

The St. John Lutheran Church in Ellenwood was built in 1879 and dedicated in October of that year after less than two months of construction time. The land was purchased from the Santa Fe Railroad for $1.00 while the cost of the frame church was around $800.00. Wilhelm Schrepel was general contractor and charged $150.00 for his labor. Painting of the exterior waited until 1881. By the turn of the century, the church facility was proving inadequate. On September 9, 1900, the congregation resolved to build a new church.

Fredrich (F.H.) Schrepel was in need of a barn, and after discussion and agreement, he bought the barn. The shingles, roof and cupola were removed; the sides and ends were laid on the ground and moved about a half of a mile north, northwest on a skid. When the church was reconstructed as a barn, both sides of the church were put together on the same side. This made the barn twice as long as the original church, being connected in the center with two large sliding doors. The other side was made new. One end was the front of the church and the other end was used for the back with a lean-to added on to make a larger barn. When the roof, shingles and cupola were on, F.H. had a new, yet rebuilt barn. Although it was larger, it still looked like a church, with the exception of the front of the church, which was now one end of the barn, where he had placed in large lettering "F.H. SCHREPEL, CHEYENNE VALLEY STOCK FARM, PERCHERON HORSES".

F.H. Schrepel had a large horse operation consisting of prize-winning stallions and mares. He transported and exhibited his Percherons at the Chicago Worlds Fair in 1891, St. Louis, Missouri Worlds Fair in 1910, and numerous other horse shows. During the winter, prospective buyers would come to see his horses. F.H. would show his horses inside the barn, using the light coming in from the large old church windows as natural lighting.

The barn was also used for auctions and family reunions. Hay was kept in the north end, as there was no loft. The hay was put in a large pile to feed later. The kids had fun swinging from a rope into the hay.

BARTON COUNTY, continued

In the early 1930's the Industrial Revolution was taking over farms. It was the end of a saga, but F.H. had sold most of his draft horses years earlier. F.H. sold his land, home and barn to a cousin and moved into a small house in town, where he died in 1931. The Polhmans, to whom F.H. had sold his farm, used the barn for a dairy for ten years. The barn was torn down in 1939 or 1940. The lumber was used to build a fairly good-sized shed where the barn stood. The shed is still in use today.

Trego County
Martha Knudsen

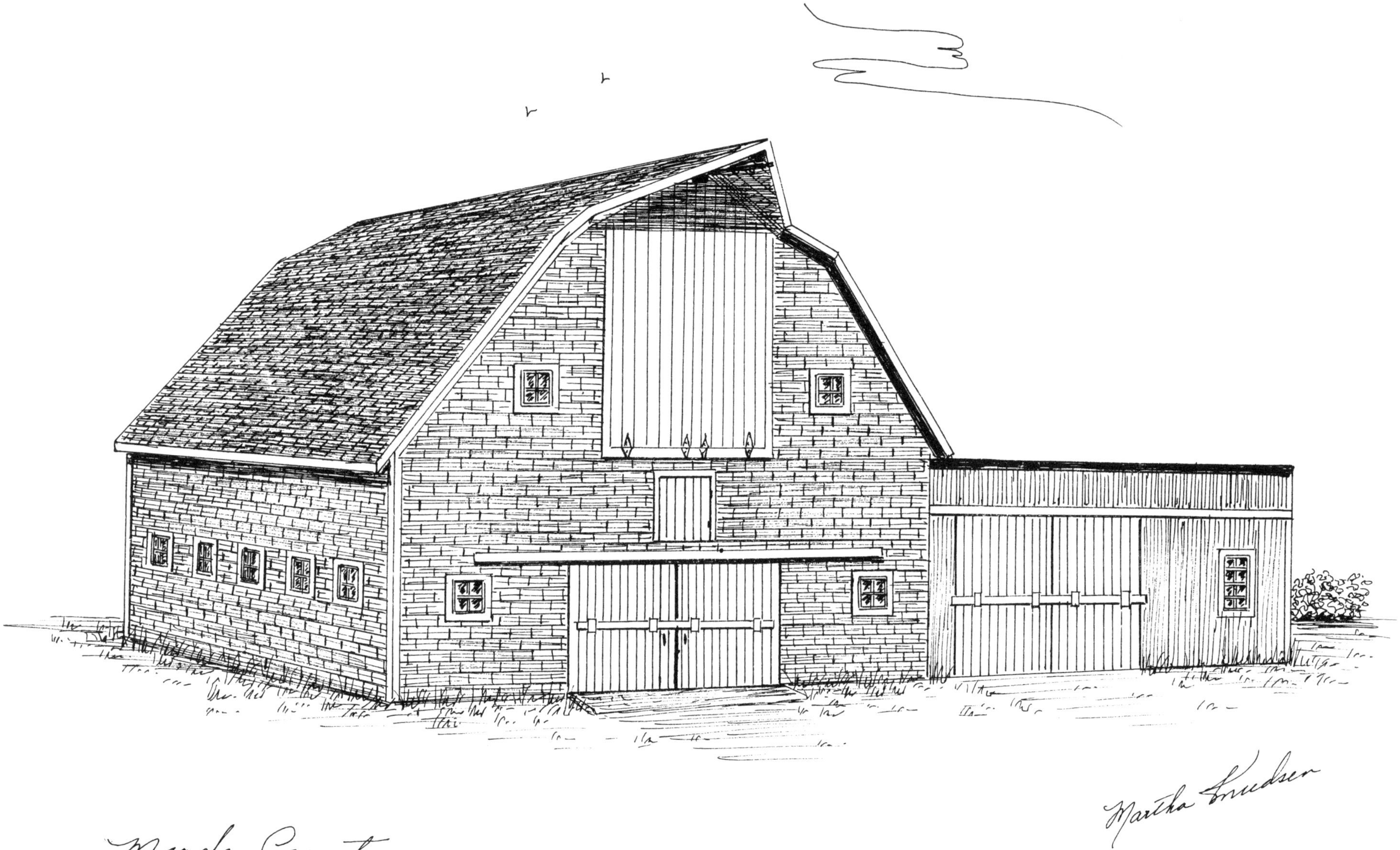

Meade County

This small barn was built around 1910, for horses and milk cows. It is one of a few barns that were shingled all over.

It has been a landmark for years as it sits halfway between Meade and Englewood.

Pratt County

Carl Hertlein and his younger brother, Ed, built the barn in 1939 or 1940, for cattle, maybe a few horses, and hay and grain storage.

Carl was very observant and noticed how the sun's shadow fell on the barn and looking at his watch saw it was 10 o'clock, so he marked the shadow's place on the barn. By watching the shadows, he came up with the rest of the numerals. The numerals could be seen from the field, so they may have used the barn to keep track of the time while out working in the field. It is thought the time dial was made in May or October.

This barn is located southwest of Pratt.

Ellis County
Martha Knudsen

This barn has been gone for many years, but the history of the ranch and the clock was so interesting, that I thought it was worth including in the book. I'm sure there were many more interesting ranches of this nature that we will never know about.

The land the barn was built on first belonged to the Kansas Pacific Railroad Company. In October 1872, they agreed to sell it for $2.00 an acre. In March of 1873, the deed for the land was given to George Grant, founder of Victoria. Mr. Grant gave the deed to William Ratchliff in June 1875, and on April 30, 1879, it was deeded to Rubbens Battell, a wealthy English nobleman who owned large estates in almost every country in the world. He named Jim Beken as his manager.

He owned this land for forty-one years, but only once did he see it. He owned 5,600 acres in Ellis County. In addition, he had large, beautiful homes in Connecticut, New York, Indiana and also owned real estate in Florida, Virginia, Iowa, and Illinois. Having so much property was one of the reasons he never got back to Ellis County again.

The Beken ranch was noted far and wide because it was stocked with prize-winning cattle, including Aberdeen Angus, which had been imported from England when the land was still owned by George Grant.

The barn was built in 1880, according to the Elizabethan style of English architecture. Mr. A.L. Hatchkis installed the clock. The Seth Thomas Clock Company of Thomaston, Connecticut made the clock. It was considered the biggest clock in western Kansas. At the age of sixty years, it still was an excellent timekeeper, and could be heard striking the time as far away as five miles.

The clock was mounted on a steel stand and the timepiece was driven by a one hundred thirty-pound weight with a 3/16-inch steel cable. It has a device that strikes the hour according to the time. It was controlled by a big thirty-pound hammer, which strikes on a large bell. It is driven by a three hundred-pound weight on a steel cable, and is controlled by a governor. The pendulum is about four feet long and weighs over one hundred pounds.

The larger gears were made of steel, the smaller gears of bronze, and the numbers on the dial of bronze. These numbers are about one-half inch thick. The clock has minute, second and hour hands that are made of poplar wood. The clock is wound with a crank about every eight days.

When completed, the clock weighed about one ton and cost, with installation, $4,837.00

The barn was destroyed by a tornado in 1901. After the storm passed, the clock was found among the debris, one hundred fifty yards from the foundation. It was hardly damaged, except for a few minor damages, which were easily repaired. The clock was placed on the new barn that was rebuilt that same year.

The house on the ranch was a huge stone structure, having two wings, each three stories high. One wing was built in 1880, and the other in 1881.

A screened veranda enclosed the northeast wing on all but part of the west side. This wing included a stairway, two hallways, a playroom, two bedrooms, a guest bedroom, a completely furnished bathroom and a parlor. The guestroom and the parlor had beautifully built-in fireplaces with mantels. The parlor had three large bay windows on the east side. There were also windows on the north and south sides of the room. The woodwork in the room was of walnut.

The southwest wing included three stairways, a hallway that extended the length of the house, four bedrooms, a washroom, storeroom, kitchen, dining room, and a large
ELLIS COUNTY BARN, continued

living room. On the east side of the living room was a doorway that led into a sunroom. Years later, the room was removed and a door to the outside put in its place. On the south side of the dining room was a door that led into an enclosed porch. This porch had two entrances that opened onto the veranda.

In all, there were fifty-six doors in the house.

ELLIS COUNTY BARN, continued

There also was a large orchard with many kinds of fruit trees. An underground room was built about thirty feet from the house in case of storms and it came in handy several times, including when the barn was destroyed in 1901.

The ranch was very large and had many buildings including a bunkhouse. A large corral was constructed of double walled stone with an arched rock on top. This corral enclosed a building, which was separated into compartments that were used as cow sheds and stalls for feeding horses. It was three hundred feet long and was built along the whole north end and part of the west side of the corral. The barn was on the west side.

The ranch was sold several years after World War I, when prices were at their highest.

In 1930-31, the old ranch house was torn down and a new house built about a quarter of a mile southwest of the original sight.

The clock was moved to a special building built for it in 1936, about 130 feet west of the house. A tractor was used to hoist the clock from the cupola onto a wagon. A team of horses was then used to haul it to the special building where it could be seen again. Tourists and neighbors often came to look at the clock and see how all the devices worked.

Eight months after the clock was removed, the barn was once again destroyed by a tornado.

To date, the fate of the clock is unknown as it was sold and no one knew where it is, or if it is still in working condition, or in parts.

Butler County

WILCOX BARN OR FORT

Horace Wilcox was born in Connecticut in 1821 and grew up in a dirt-poor environment that probably contributed greatly to his "me against the world" philosophy. In 1849, with nothing to lose, he joined the Great California Gold Rush. Unlike most of the 49'ers, he struck it rich, and bought 400 acres of prime Illinois land and temporarily settled down for the good life. That may have been the last time he tried to get along with his neighbors.

Early in the 1850's, he was helping his neighbor with his threshing. When they went in to noon dinner, he spied a young woman who was helping with the cooking. He learned her name was "Olivia", and so began the shortest courtship around. That evening they walked to town and were married. After four children, Horace decided to become a cattle baron, so he loaded up the family and moved to Denton, Texas.

The storm clouds of the Civil War were gathering on the horizon, and Horace was a "blue bellied Yankee" in a Confederate state. When the Civil War broke out in 1861, a committee of Horace's more charitable neighbors met with him and advised him to "get out of Texas in 24 hours" or a less charitable group would form to give him a suspended sentence from the nearest tree. Horace wasn't stupid. He told his wife to load the wagon and children and he'd be waiting for them in what is now Oklahoma. He then saddled his best horse and headed north at a dead run. The vigilantes were close behind, but he outran them.

He was waiting when his family reached Oklahoma. So were the vigilantes, but he didn't know that. When one of his youngsters spotted him and yelled "There's daddy!", the chase was on. It didn't last long, the vigilance committee decided he was no longer a threat to Texas or the Confederacy. Horace and his family were reunited and they headed north. Where Wichita now stands, they came upon a well-known Indian trail leading northeast and followed it.

WILCOX BARN OR FORT, continued

In northwestern Butler County, they went through what would later be known as "Oil Valley" and crossed the Whitewater River a mile south of where he eventually would build his fort. As they were passing by, he said that he would settle there one day. Horace and his family went on to Illinois, but as he had not forgotten his dream of becoming a cattle baron, he sold out in 1866 and returned to the to the Whitewater River, where they built a log cabin.

Two years later, he quarried the stone and built his fort. The fort really was a three-story barn. The lower level had eleven strategically placed gun ports, each with a steel cover, and a door through which he could ride at a full gallop. The second floor was the barn proper. The third story had a large finished room at one end. Records show that Masonic Lodge meetings were held there as early as 1868. All the massive timbers in the fort were hand-hewn. Horace had to buy all the hardware at Junction City, 100 miles north.

By 1870, Horace's cattle empire was beginning to show promise, so he built a suitable house. The fort was a big-enough do-it-yourself project that he hired some Swedish stone masons in the area.

There was still trouble in Horace's version of paradise. In addition to his cattle, he had several buffalo calves because he liked them. One day, he discovered that one of the calves had been shot, and he assumed it had to have been done by his neighbor. True to form, Horace strapped on his guns and rode over to shoot his neighbor. It took some fast-talking to convince Horace that he had the wrong man.

Horace and Olivia had high hopes of developing a large cattle herd. Together with their two oldest children, who had married in 1869 and settled in Harvey County, they had dreams of forming a vast triangle of grazing ground for their hundreds of cattle. With the Chisholm Trail only a few miles to the west, their market outlet would be assured. It was a dream that lasted only a few years.

Settlers were moving into the upper Whitewater area, stringing fences and plowing up the prairie, so Horace decided to head for open range in Oklahoma. He might have found it except he used a couple of stray Indians for target practice and barely made it back to Kansas with his scalp.

Horace's next project was to turn his fort/barn into a mill. He built a huge "wind-wheel", a sort of ground level windmill for power. It worked, but made a terrible racket.

The other settlers had tried to get him to donate land for a school. He refused, so they built one just outside his property line. Shortly thereafter, the school was the site of a big revival meeting. Not liking this, Horace turned on his wind-wheel. When the noise drowned out the preaching, the revival stopped long enough for those in attendance to go over and chain the wheel. A sudden gust of wind caught the device, turning it over and smashing it.

Apparently, Horace figured someone up there was telling him something, so he built the Whitewater Center Church a mile or more north in Marion County.

Horace died in 1888, and a tall monument marks his grave in a nearby cemetery. His widow ran the farm for awhile, but moved to California where she died in 1904.

Barns this old made of stone, timber and wooden pegs are not moved every day. Mr. Bob Green, an attorney from ElDorado, undertook this task in 1983, and moved the barn to a site east of ElDorado and rebuilt the barn to the same dimensions and thickness. "To preserve something of historical value" was his motivation.

Johnson County
Martha Knudsen

The barn is on the list that the wrecking crew has had in hand.

On the northwest edge of Olathe, homes are being built on several sections of land, and this barn happens to be on one of them. It was a beautiful barn in its day, but it looks like it has suffered many years of neglect.

It is twelve sided and 75 feet tall and 75 feet in diameter. It was built with a drive-through the center of the barn. Horse stalls were on one side with granaries separating them from the drive-through. The other side was for cattle.

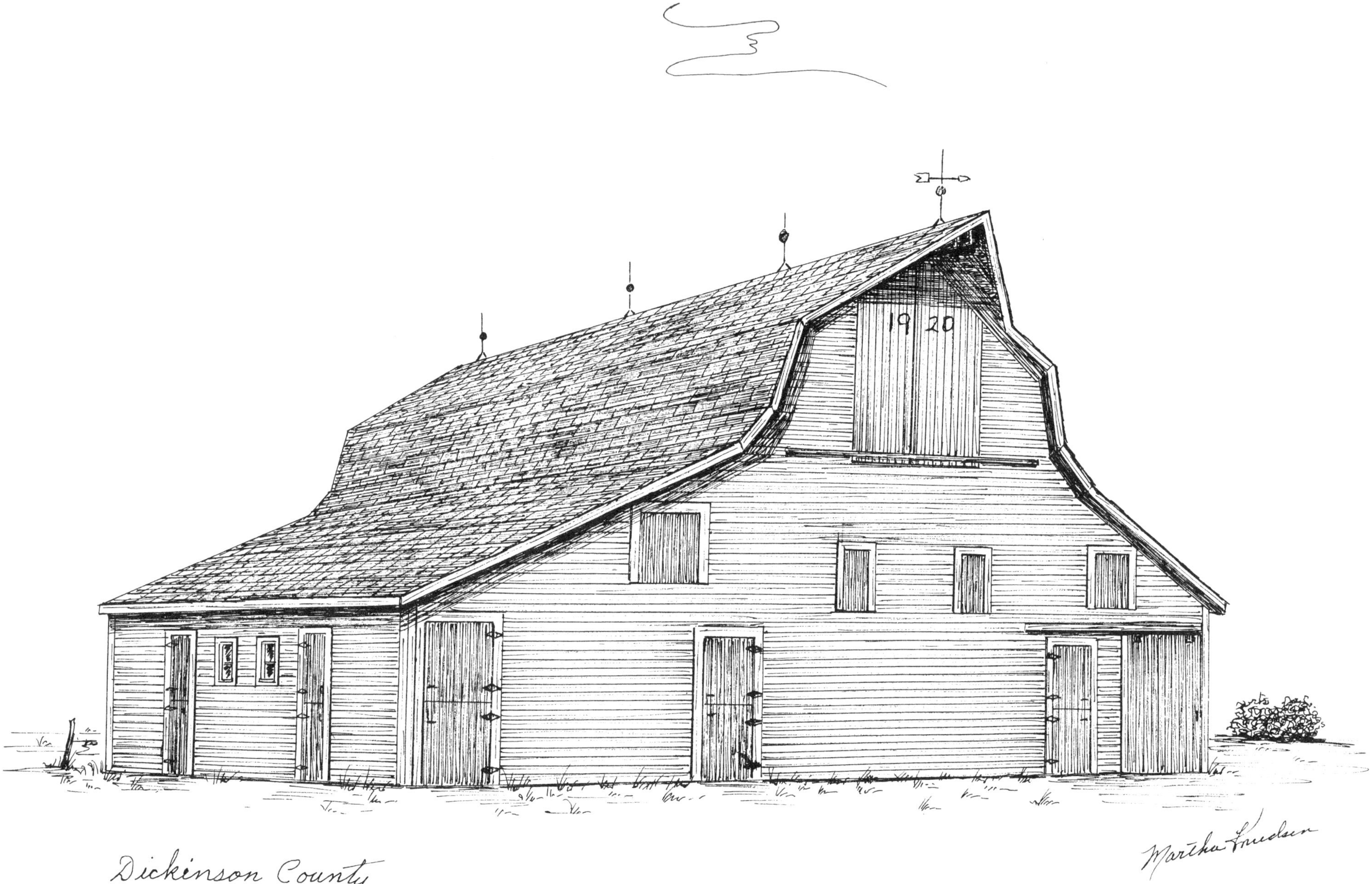

Dickinson County

The Woods barn southwest of Carlton, in Dickinson County was built in 1919 by W.L. Woods and painted the first time in 1920. Mr. Woods came to Kansas at the age of 10 with his parents in a lumber wagon from Iowa. The family settled southwest of Carlton. W.L. "Will" Woods married Nellie Sloop in 1898, and started farming and raising livestock. It wasn't until after the birth of their sixth child that Will decided to spend the money and built a barn that would become the hub of the livestock operation.

The center of the barn was built to hold loose hay and was filled with slings. Hay moved from the hayrack up under the eaves of the barn, then traveled back across the barn on metal tracks before being dumped in the center of the barn. Also in the center of the barn was an oats bin and cream separator room

The entire south side of the barn was built for horses, with room for ten head at mangers, or five box stalls. The north side of the barn included stanchions for nine milk cows near the cream separating room. Skin milk was fed to the baby calves inside the barn. The barn was still in good repair long after the horses and milk cows were no longer profitable in the farming operation.

Mr. Woods died in 1946, and his daughter, Erma Van Meter, and family moved to the farm in 1949. The barn was tinned in the 1930s. On the outside south wall of the barn are many handwritten notes of breeding and calving records made in the 1930s and 1940s by W.L. Woods.

The barn is no longer used and is showing signs of the times with loose and missing tin and loose and missing barn boards.

Clark County

This ranch in northern Clark County, originally the Leustrum/Carlson Ranch, dates from the latter 1870s.

The barn was actually a schoolhouse that was changed into a barn when additions were added all around the school to make up the new structure.

Though the date of the building of the barn is not known, a map of Clark County in 1886, shows School District 5 at the location of the barn as it is located today. From May of 1886, to November of 1890, the town of Vanhem was located here. Though proof is lacking, it is felt that the school dates from 1886 to 1890, with the barn built around the school house shortly after that as School District 5 does not show up on a 1909 map.

The barn is showing its age now, but is very different from many of the barns you normally see around today.

Jefferson County
Martha Knudsen

When Walter Kay purchased the farm, he decided to have a round barn built. This barn was started on December 10, 1910. Lumber for the project arrived at Oskalossa on two extra length flat cars. Carpenters who worked on the barn were Jack Slawson, George Sands, and John Terhune.

The center of the barn was built first. During later years, the framework began to do what some other round barns had done. Many times, round barns will twist in a counter-clockwise direction. In order to counteract this twisting, an outer circular barn, or shed, was added in 1922.

The center section has a diameter of 32 feet and a height of 50 feet. The added portion of the barn is 16 feet wide with two large drive-in doors and 32 windows.

The farm is located on a hill two miles east of town. The farm was later sold to Otto Ickes and then to Duane King. There are different owners on the farm now. Today, there is only a small section of the outer shed left and like the majority of barns today, it is starting to show it's age.

Harvey County
Martha Knudsen

This barn was built in 1912. The owners have kept up with the repairs and painting so that the barn is in good condition. The farmyard is also very neat.

Harvey County

Martha Knudsen

The dairy barn, pictured in the top left, was built for Dr. Arthur E. Hertzler, M.D. in the 1920's. The barn was a show place, built with the very best of materials and dairy equipment for that time. Dr. Hertzler didn't live there, but had a manager live on the place. The dairy milked Holstein cows.

The other barn was for the horses used in farming and was probably built around the same time.

The Mueller's had a dairy there from 1941 to 1969.

At this time, someone lives there, but I don't think the barns have been used for years. They are in good shape considering their age and the fact that they are not being used.

Elk County

These barns were built for Sam Hanna in 1893 or 1894. Cypress and rock were used to make strong, sturdy barns that would last for many years.

The cattle barn, pictured top left, was for Mr. Hanna's imported Shorthorn cattle from Scotland. There were box stalls for the cattle and the hayloft contained grain bins along with the loose hay. There were slides from the bins to feed the cattle in the different stalls below as well as shafts for hay distribution. The barn had two cupolas at one time.

The horse barn was for Mr. Hanna's imported Percherons from Scotland. There were horse stalls below and hay and grain bins in the loft for distribution just like in the cattle barn. A pole shed was added later. Mr. Hanna also raised Hampshire hogs.

Behind the barn on the hill was a cistern that drew water from the creek below into a large concrete tank. From there, the water was sent down to the barns.

Mr. Hanna also had many acres of grassland in western Elk County.

The barns are in good condition for their age. The larger barn is still used at times.

Grant County

Cowley County

Late 1800's

ABOUT THE AUTHOR

Martha Knudsen graduated from Moundridge High School and went on to Northwestern University in Chicago, where she graduated as a Dental Hygienist. She worked 40 years in dental offices in Memphis, Tennessee, Wichita and Newton, Kansas before retiring in 1986.

After retiring, she worked in oils for two years before changing to pen and ink. A self-taught artist, she started with barns and windmills because they had straight lines and very little shading. Over the next two years, she developed more of an architectural style of drawing. This led to a series of Kansas things: stone houses; schoolhouses; courthouses; churches; windmills and two sets of barn notes.

In 1991, her Kansas Barn Calendar came about. She started with four different barns from different counties and then set out to find eight more barns from an additional eight counties. This year marks her tenth calendar, every barn different. In 1993, she developed a Kansas Barn Book with the first four calendars and twenty other barns and as many histories as she could find. With this book, plus the first one, she has depicted at least one barn from every county.

Martha has exhibited her work at Century II in Wichita, Nations Bank (Bank IV) in Newton, The Carriage Factory Gallery in Newton; The Newton Public Library and at the Harvey County and Kansas State Fairs.

In addition, she attends numerous arts and craft shows throughout the state and has consigned her work at stores featuring Kansas products.

Martha was featured as one of "Hatteberg's People" in 1993 and was included in his book "More Kansas People".

She lives with her husband on a farm near Walton. They have two daughters and two granddaughters.

She can be reached at: Martha Knudsen, 4028 N.E. 48th St., Newton, KS 67114